Automotive Repair Case Studies

(Diagnostic Strategies of Modern Automotive Systems)

By Mandy Concepcion

All charts, photos, and signal waveform captures were taken from the author's file library. This book was written without the sponsoring of any one particular company or organization. No endorsements are made or implied. Any reference to a company or organization is made purely for sake of information.

The information, schematics diagrams, documentation, and other material in this book are provided "as is", without warranty of any kind. No warranty can be made to the testing procedures contained in this book for accuracy or completeness. In no event shall the publisher or author be liable for direct, indirect, incidental, or consequential damages in connection with, or arising out of the performance or other use of the information or materials contained in this book. The acceptance of this manual is conditional on the acceptance of this disclaimer.

Preface

In this section, we'll take a look at automotive diagnostics in action. An effort has been made to look at problems in different ways, in each of the examples. Although there are many ways to perform the same task, the idea here is to show the technician or avid DIY mechanic the different ways to go about diagnosing automobiles.

Special attention is given to specific systems and different makes and models. The different real life diagnostic cases are explained from narrated perspective to make learning easier. Hopefully you find this section enlightening and productive. Enjoy your readings.

Developed in the USA

Automotive Repair Case Studies

Diagnostic Strategies of Modern Automotive Systems (Vol 9)

Table of Contents

* - Audi data bus signal recognition (exposes the intricacies of diagnosing vehicle networks and how computers talk to each other.)

* - Cadi idle re-learn (explains the importance of module re-learn procedure, which is done by re-adapting the ECM to a new sensor.)

* - Case of the EVAPs (these emission systems are difficult to diagnose, due to their complexities.)

* - Computer Data Lines (scan tools talk to the different engine modules or computers through the data line or bus. See how to diagnose this type of problem.)

* - Faulty EGR operation (the EGR is in charge of lowering combustion temperatures. But issue with this system can cause pinging, performance, misfire and countless other issues.)

* - Lean (dirty) MAF (the lean condition comprises about 60% of all engine performance issues. Learn to deal with this situation.)

* - The case of the low volume (Fuel pumps deliver both pressure and volume. If one of these is missing then the engine has problems.)

* - Unsynchronized CAM & CRK signals (CAM and CRK signal synchronization is needed for the engine to start.)

* - Wrong MAP reading (The manifold air pressure is a main input to the ECM. See how this sensor creates havoc with the engine.)

* - The Cadi's dual crank affair (this Cadillac's engine control system has dual crank sensors. Learn to diagnose these systems.)

* - Analytical misfire code (Misfires are difficult to diagnose and this case shows precisely that.)

* - The misfire ghost (A case of hard to find misfire.)

****

CASE STUDIES

Section 1 - AUDI DATA BUS SIGNAL RECOGNITION

This was a 1999 Audi A4 Quatro with a 1.8L TC engine that was having a hard time communicating with the scanner. The dash lights for the ABS, SRS and brakes were on all the time. The engine always started, but the customer wanted to solve the dash lights problem. Communication was never attained with the ECM module, ABS, SRS or with some of the other less common modules. The instrument cluster module however was communicating, with only an ABS code for "No communication". One of the first diagnostic questions was; why were the other modules not communicating with the scanner? Why weren't any other codes present for the other modules? And what was the system data bus signal parameters supposed to be?

At this point I knew that the key to solving this vehicle's dash lights problem was to properly diagnose the data bus fault. In modern automobiles, the data bus is the communication highway through where all present modules communicate. A faulty or shorted module can bring down the data bus and the rest of the modules with it.

I started the diagnostics by base lining the system, by simply taking a reading before the diagnostics process is started. This way the results can be compared once the repair is done.

I scoped the data bus at the DLC connector PIN # 7 to capture the waveform for later comparison. See Fig 1. The VW/AUDI vehicle line uses what's called the K-LINE to communicate with the scan tool. This K-Line however is not used for internal module to module communications. The vehicle's modules do not care whether this K-Line is down or not. It is only used for scanner diagnostics communications. All modules use a different set of wires for inter-module communication.

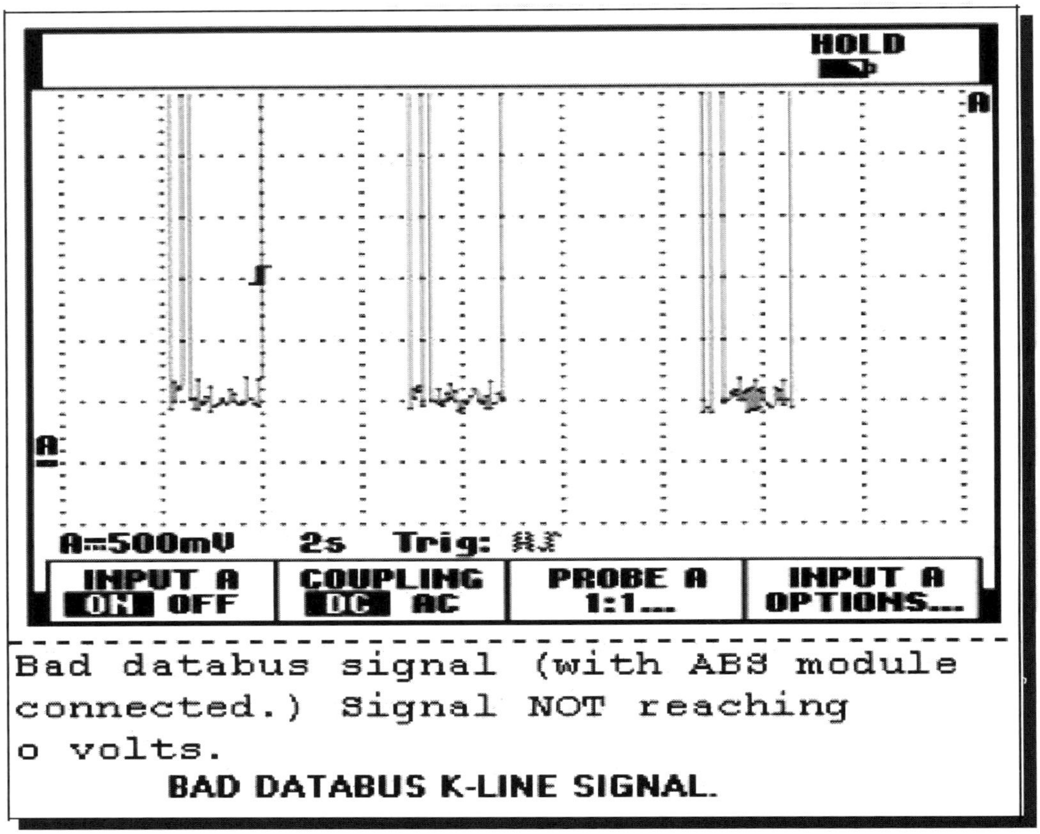

Fig 1 - Defective K-Line data bus signal.

Fig. 1 shows the waveform capture of the bad diagnostics K-Line. It shows the K-Line diagnostics data bus signal while the scanner was trying to communicate with the ECM. The three digital pulses belong to the scan tool grounding the K-Line. Each of the three digital pulses is composed of a short pulse plus a long pulse. This is the coded ID for the ECM module. Every time a different module was selected at the scan tool menu, the resulting communication ID pulses where different. Each module present in the vehicle has a different ID code that the scan tool used to access them on the diagnostics K-Line.

In Fig 1, we can see that the scan tool was grounding the K-Line. However, a careful observationof the waveform capture also shows that the toggling data bus digital signal never reached the actual 0 volt orwas never completely grounded. It was always 500 mV above ground. Was this enough for the ECM not to recognize thedigital data signal? Was the threshold or point of data pulse recognition lower than 500 mV? Obviously yes, since thescanner was not able to establish communication.

But what could cause this data signal not to reach full ground potential? The answer was a partial short to battery voltage, but no enough to be a full short.

In data bus systems like this one, an electrical problem with one of the modules, even a minor one, is enough to bring down the K-Line or any data bus in general.

The answer to this problem was to start disconnecting modules until the partial voltage short went away. I decided not to touch the engine module (ECM), so I started with the most easily accessible module. I disconnected the ABS module and to my amazement the short went away. I then tried to re-establish communications with the SRS, ECM and the rest of the modules and was successful in doing so. I had found the source of the data bus (K-line) short. A new ABS module took care of the problem. I also performed all the re-coding, "basic settings" procedures and erased all faulty codes. It is worth noting that the instrument cluster module was able to communicate, as well as the radio. From the looks of it, the digital recognition threshold on these modules is slightly higher, since communications was always established.

NOTE: It is common to see a shorted to power K-Line on these vehicles. The reason for it is that, on newer vehicles, the radio has diagnostics capabilities and also communicates on the K-Line. When the radios are swapped with older ones, by audio stores, the electrical pin-out is not the same. The older radio actually shorts the K Line to power and no scan tool communication is possible. This however has no effect on the inter module communication data bus and the vehicle will operate normally.

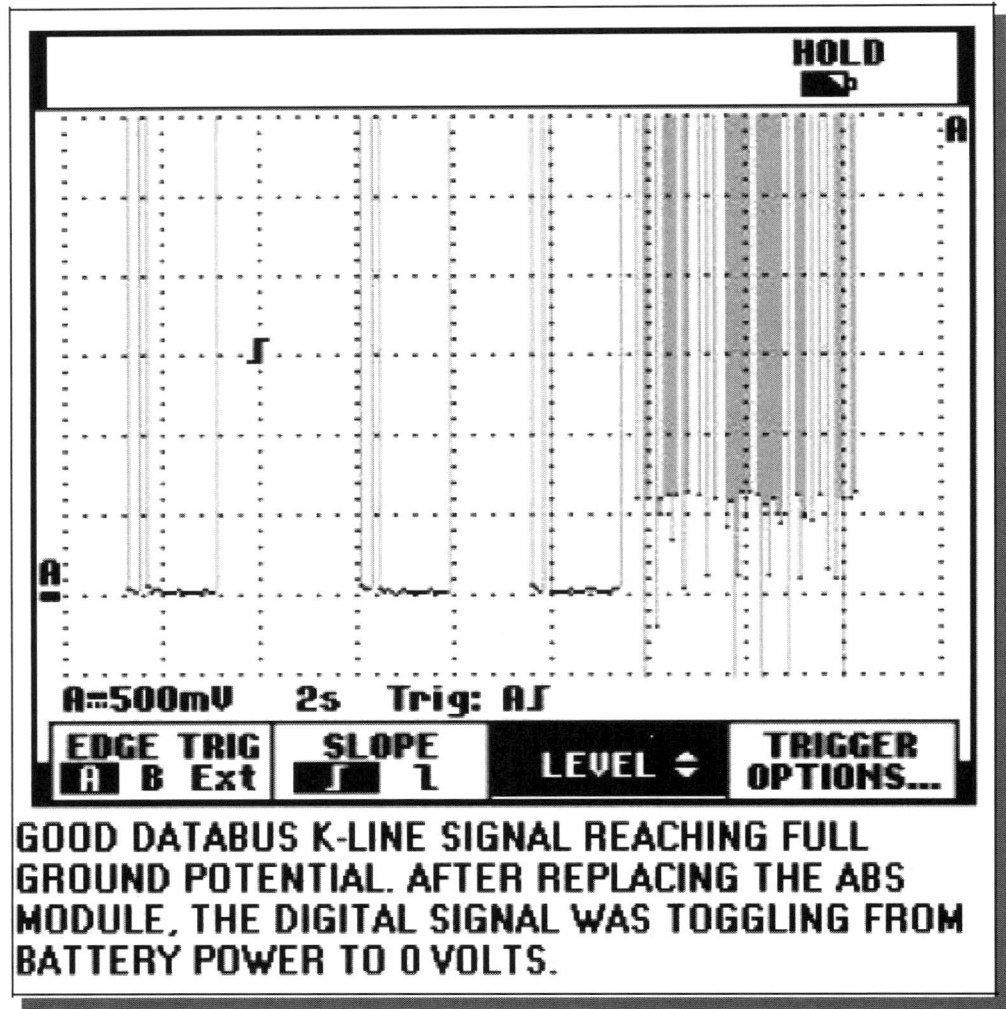

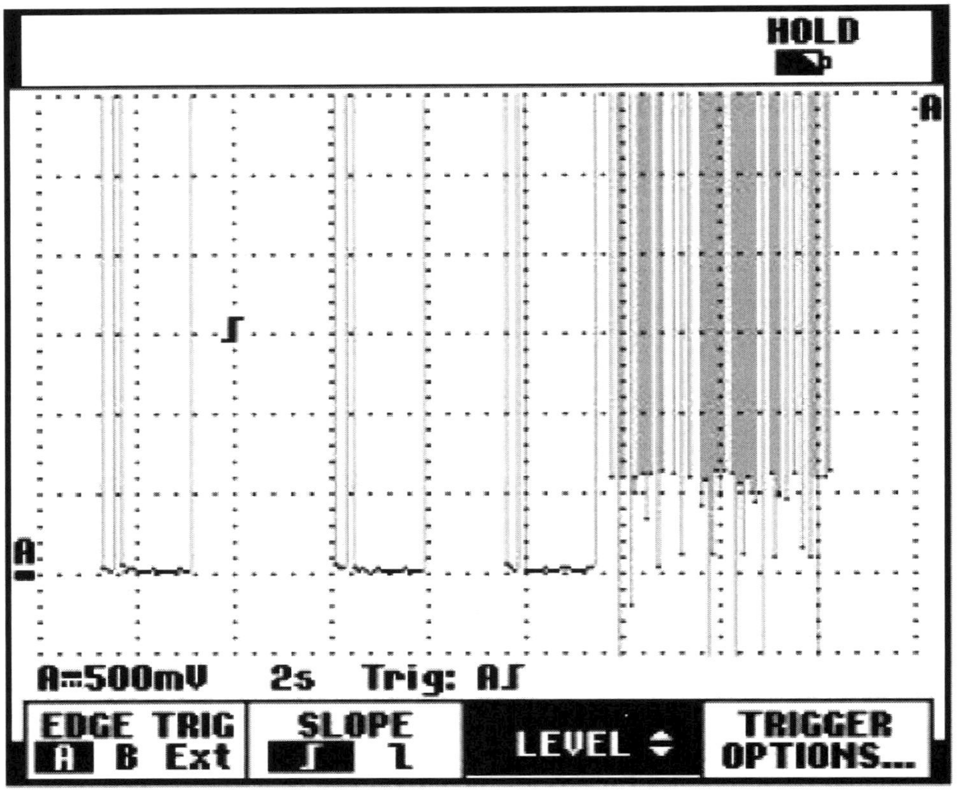

GOOD DATABUS K-LINE SIGNAL REACHING FULL GROUND POTENTIAL. AFTER REPLACING THE ABS MODULE, THE DIGITAL SIGNAL WAS TOGGLING FROM BATTERY POWER TO 0 VOLTS.

Fig 2 - Proper K-Line database signal after ABS module replacement. The first 3 short-long pulses is the scanner establishing communication with the ECM module. The higher frequency pulses that come after is the actual serial data communications already established.

This case brings the ever-present main rule of diagnostics. Know the system being worked on. Having the right information is vital in modern vehicle diagnostics.

****

Section 2 - CADILLAC IDLE RE-LEARN

This vehicle was a 93 Cadillac Seville STS with a 4.6 (vin 9). It came in with a code P080-Idle Re-learn not complete. The customer's complaint was hard shifting and check engine light on. The idle seemed to be higher than normal at about 1050 RPM. The customer also feared that the transmission was pretty much done (defective).The first step was to see if the IAC motor was operational. This IAC system was a throttle plate actuator. The actuator acts directly on the throttle lever, which also affects the TPS. This is typical of Cadillac. Therefore, a faulty IAC also affected the transmission shifts points, since the TPS also affects transmission shifting. I started by commanding the idle up-and-down. However, in this system the fine idle is controlled by timing. The ECM simply advances or retards timing accordingly to control the idle speed and there wasn't an actual IAC bi-directional control ability from the Tech-2 in this system. This was one of the early 4.6L Northstar engine years.

After performing further checks, I found out that the IAC motor was completely stuck. It wasn't moving at all, even when loading (steering, A/C on, etc) or starting the engine. Since the IAC motor was stuck in one position, it was preventing the ECM from performing an idle adaptive re-learn. This IAC control system moves the throttle plates to control the idle (course idle adjustment) and the TPS readings were off specs. As a consequence, the transmission wasn't shifting properly and the idle speed was also skewed. Modern electronically controlled transmissions use the TPS as a main TCM input signal for throttle opening and rate of change. The TCM uses this signal for proper shifting. The customer actually came to the shop with a hard shifting problem, not an idle complaint, although it was slightly higher than normal.

The key factor in this particular case is not to get confused by this particular vehicle's idle speed control operation. Since the idle speed is partly controlled by timing (fine idle adjustment) a scanner idle actuation command to verify the IAC is simply not possible. A visual inspection and the fact that the idle speed and TPS signal readings were higher than normal revealed the stuck IAC motor. In the world of automotive diagnostics nothing is as it seems sometimes. Just because a code is present for component A doesn't mean that component A is faulty. In this case, the transmission was not at fault and a new IAC motor made the transmission shift like charm again, after the idle re-learn procedure was done.

****

Section 3 - CASE OF THE EVAPs

1998 Chevy Venture 3.4L.

This vehicle came in with an EVAP code P0440-general EVAP system leak. No other codes were present and the engine was running fine. The fuel cap also tested well. The diagnostics system was OBD II with full bi-directional controls. I started by connecting the Tech-2 and de-coding the ECM (code P0440). This GM-EVAP system uses a purge and a vent valve to control the vacuum to the fuel tank and then monitor the fuel tank pressure sensor for vacuum drops. I performed an EVAP seal test using the Tech-2. By simply closing the vent valve, commanding the purge valve to 100%and then monitoring the fuel tank pressure sensor for a vacuum loss, it was easy to determine that indeed there was a leak. The question was where was it? The answer was to perform this test a couple of times while isolating the different components.

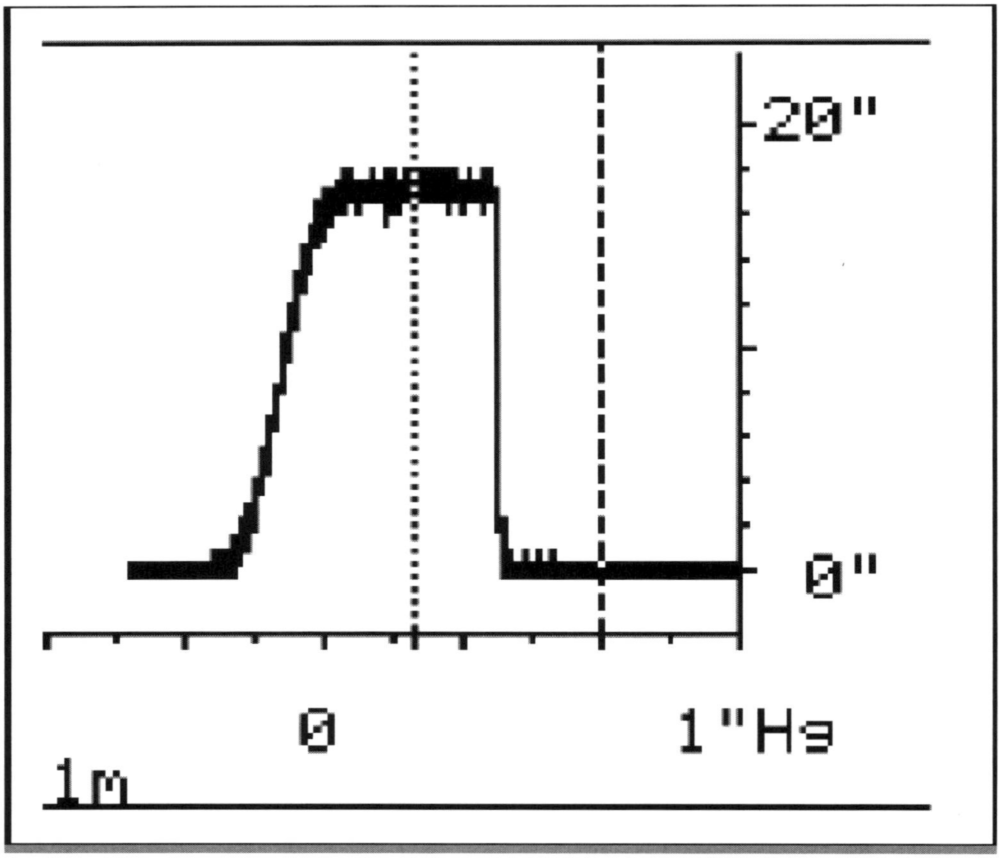

Fig 1 - Seal test commanded, the waveform shows the vacuum going up then leaking down.

I started by testing the purge valve, vent valve and canister under the vehicle for any leaks. Each time I did the test, one of those components was isolated. But the vacuum was still dropping each time. My last resort was to isolate the fuel tank by disconnecting it form the canister and plugging the port. This however would render the fuel tank pressure sensor useless, since it was placed on top of the canister. So I used an electronic pressure/vacuum transducer to monitor the EVAP vacuum while commanding the EVAP seal test with the Tech-2. When the fuel tank went out of the picture, I could see then that the vacuum was going up to 17 ″ Hg and stayed there.

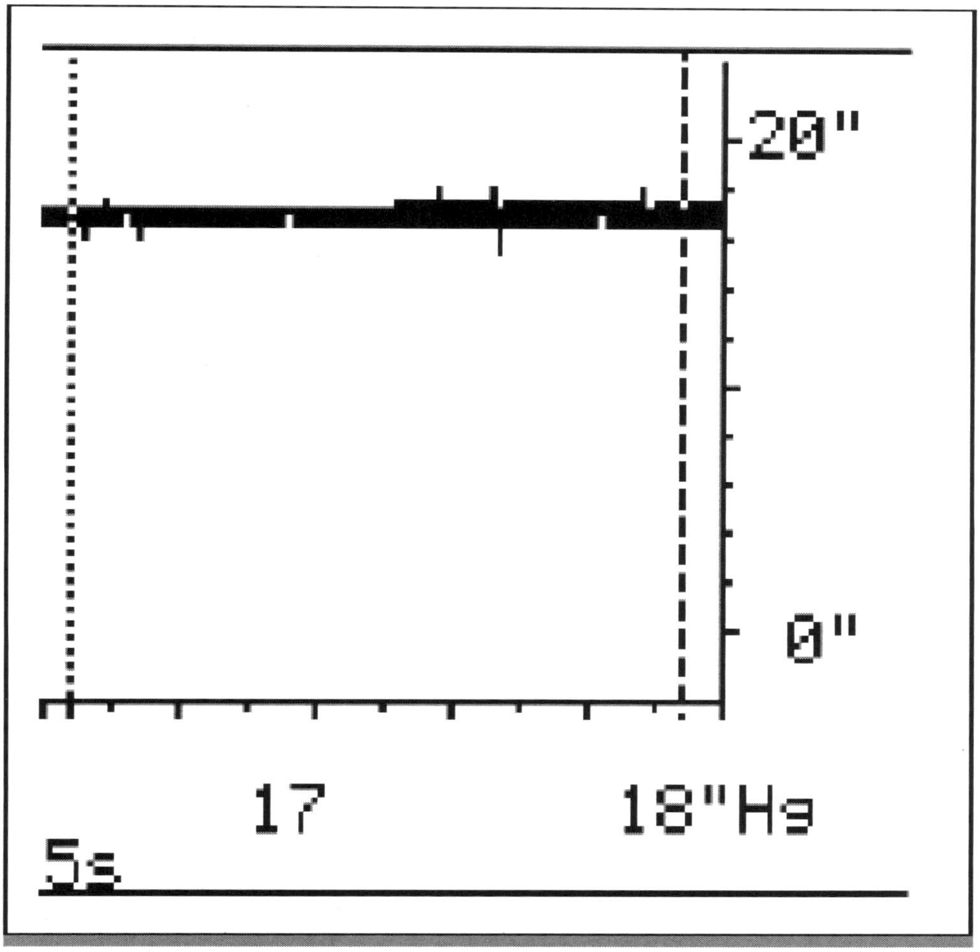

Fig 2 - EVAP vacuum holding steady with fuel tank disconnected from system(eliminated).

This was the indication I needed to reach my final repair decision. The plastic fuel tank was simply leaking. Sure enough a complete replacement of all the O-Rings, hoses and seals solved the problem. It is important to note that after the repairs were done (before mounting the tank) a vacuum test was done on the tank using a hand vacuum pump. Therefore, eliminating the possibility of a faulty repair after the tank was in place. Better to be careful than sorry.

****

Section 4 - COMPUTER DATA LINES

This was a 1997 Cadillac De Ville with a 4.6L (Vin Y) engine that came in with an inoperative CVRSS with ELC(Continuously Variable Road Sensing Suspension/Electronic Level Control). The vehicle simply wouldn't go up at the rear and the suspension was stiff all the time. I started my diagnostics routine with a simple code scanning to see if there were any relevant codes. There was only a no communications with the CVRSS module message in the IPC (Instrument Panel Cluster), but no codes. This was my first clue.

Note: In this particular system, as in most late Cadillacs, the CVRSS module stores its codes in the IPC memory.

After my fruitless scanning routine, I decided to go back to my Tech-2 and perform a module pinging procedure. TheTech-2 has a feature at the beginning of the menu that allows the technician to call-up or ping one or all modules. This operation revealed that the CVRSS module was not present in the vehicle, but why? I knew this vehicle came equipped with one. The multi-star data bus wiring diagram also gave that indication. So how could this module disappear? The answer was buried inside the Tech-2's software. This is how it works: The Tech-2 has what could be called a dynamic menu configuration routine. At the beginning of the Tech-2's power-up sequence, you are asked to enter the year, make and model then the scanner queries all the modules that are present in the vehicle. It is from this query that theTech-2 configures its menu of available modules. If a particular module is not present because it's faulty, it will not respond to the initial query, therefore it will not be listed on the menu. This is why the CVRSS module wasn't in the menu when I pinged all the modules. There was something wrong with this module. This in itself was a good second clue.

Note: The dynamic menu configuration can be disabled in the scanner's set-up option at the main menu. After it's been disabled, then the Tech-2 gives you the option to pick-and-choose from a wide variety of module options.

I immediately concentrated on proving the CVRSS power and grounds and they all checked good. This simply was a case of a defective CVRSS module. After ordering and replacing it, the suspension system came back to live again. The lesson here is that not only is it important to know the vehicle's system but also the test equipment as well. In our automotive technical world, the learning curve never straightens. That's the beauty of what we do or should I say the curse.

Section 5 - FAULTY EGR OPERATION

The vehicle was a 1993 Lincoln Town Car with a 4.6L eng. It came in with an emission complaint of high NOx (Oxides of Nitrogen). Oxides of Nitrogen form when the combustion process happens at very high temperatures. At over 2500 º F the nitrogen in the air combines with the oxygen to form different types of oxides. This could be caused by a couple of reasons like an overheating engine, faulty EGR system operation, carbon in the combustion chamber raising the compression ratio or any combination of factors that help raise the engine temperature.

Note: In practical applications, a certain amount of carbon in the combustion chamber will not affect the NOx factor a great deal. Unless the cylinder chambers are extremely dirty, which will also give you an engine performance problem; carbonized combustion chambers will almost never be a NOx problem.

I started the diagnostics process by performing a visual observation of the engine temperature and general working conditions. The vehicle was not overheating and it maintained proper operating temperature. This fact was verified by taking a couple of temperature readings with the infrared gun.

The 2nd step was to perform a scanner PID diagnostics. I verified that the coolant temp was operating correctly and it was. I then did a quick PID overview analysis of the A/F ratio or fuel control PIDs. The FUEL TRIMS were at acceptable levels (+2.0 %) and the O2 sensors were switching at the right frequency/amplitude.

Note: It's worth mentioning that a lean condition will definitely raise the NOx levels. The reason being is that a lean mixture will burns much hotter than a fuel rich one. The fuel acts as a form of coolant. Therefore, a rich mixture will have high CO but low NOx. This is always expected.

By doing an A/F analysis of the engine operation, I also ruled out a possible vacuum leak. Again a vacuum leak will raise the NOx levels. A vacuum leak means excess air going into the engine and excess air will lean the mixture. A lean mixture is, of course, NOx producing. The correct fuel trims and O2 sensor switching seen at the scanner ruled out any vacuum leak problem.

I then concentrated on the EGR system. I decided to also do a PID scanner analysis of the EGR system. This time however it was better to graph the PIDs because of the dynamic or changing nature of the EGR system. This is what I saw.

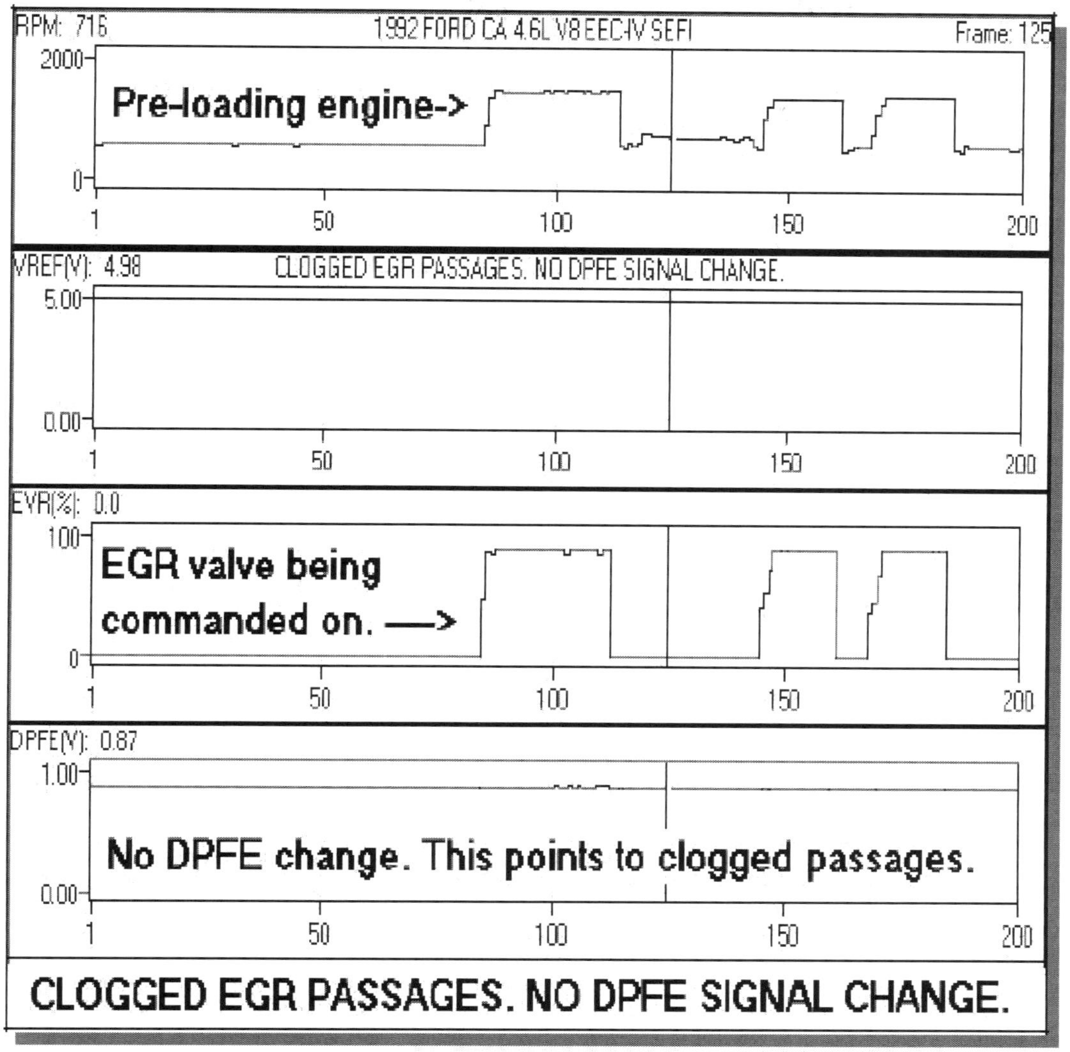

Fig 1 – Scan graph of the EGR system.

In Fig 1, we see a graphed snap shot of the EGR system while in operation. In this case a stationary engine pre-load was performed. In the upper part of the graph, notice the engine being pre-loaded (RPM going up) while the EVR (EGR vacuum regulator) percentage is also up. For the ECM to actuate the EVR a load has to be placed on the engine. The engine load factor is calculated by the ECM through the MAF, RPM and TPS sensors. In other words, the EGR will only operate while the engine is loaded up. In this case, the ECM was activating the EVR at almost 80% opening rate. This meant that the EGR valve was supposed to be getting a fair amount of vacuum and at 80% it should have opened up almost completely. However, a quick glance at the

DPFE sensor reading showed that in fact, this was not happening. The DPFE sensor is nothing more that a differential exhaust back pressure sensor. It simply compares the engine's exhaust pressure to that of the EGR tube. If any exhaust gas is flowing through the EGR valve and into the intake manifold DPFE sensor will output a differential voltage signal. In other words, the DPFE voltage should have gone up at the same time that the EVR was being commanded on or about 80% duty cycle. This analysis indicated that the EGR system was responsible for the high NOx emission failure. At this point in time, I was done with my PID scanner diagnostics. All that was left to do was to perform a couple of manual tests to arrive at a final diagnostics conclusion.

I asked myself, was this non-operational EGR system problem the result of a vacuum problem? Or was it the EGR valve itself that was stuck shut? I knew that the ECM was commanding the EVR on. So, unless the EVR to ECM electrical wiring was defective, the ECM was not the problem. I decided to go to the most accessible place to conduct my manual tests. This was at the EGR vacuum port itself. I simply connected a vacuum gauge to the EGR vacuum hose and once again did an engine pre-load. The gauge immediately went up to 5 to 6 in. Hg of vacuum. This was an indication that the EVR, electrical wiring and vacuum hoses were operating normally. Therefore, that was not the problem. The last thing I had to do was to perform a manual EGR actuation using the hand vacuum pump. By doing so, the engine speed (RPM) should decline considerably or the engine might even stall. I proceeded to the manual EGR actuation, but there was no RPM drop. I could see that the EGR diaphragm was not broken and the vacuum was pulling it up.

This could only mean one thing that the EGR passages were severely clogged and would cause the exhaust gases not to flow into the intake manifold. This was the reason for the high NOx emission levels.

The FORD 4.6L engine has a removable intake manifold elbow that when removed gives the technician access to the EGR passages. A thorough cleaning of the EGR passages corrected the problem and the NOx emission levels went back to normal. Not all EGR systems work the same but a good understanding of the system being worked on is of vital importance to the correct diagnosis and solution to the problem.

****

Section 6 - LEAN (DIRTY) MAF

This vehicle was a 1999 Mazda Protégé 1.6L. It came in with a code P-1171 O2 sensor inversion. In MAZDA terminology, it means that the O2 sensor is not switching passed 1500 RPMs with the engine warm. The vehicle was also failing for NOx (Oxides of Nitrogen) emissions.

I first performed a visual inspection, which revealed nothing, since this vehicle had fairly low miles. I then proceeded to connect the NGS scanner with MAZDA software and perform a scanner PID analysis. At first glance, it didn't reveal much. But I could see that the O2 sensor was switching fine with proper amplitude and frequency. A scan PID analysis should be performed at as many levels of vehicle operation as possible. So my next step was to do an engine pre-load and watch my PIDs. Sure enough, the O2 sensor went low to 0.03 volts. This was an indication of a very lean condition. But how could this be? The vehicle exhibited no noticeable engine performance problems. What could be causing this lean condition?

Being as is was, the problem had to do with the fuel system. So, I started to check all my fuel system basics (pressure and volume). It all checked out. So, what then? In situations like this, I always try to think like an ECM. The one thing I already knew was that the lean condition was happening off idle and with the engine loaded. What was then the main ECM load input at off-idle engine and during acceleration? This is a viable question since the fault was only happening during pre-load. This vehicle was simply going into open loop at pre-load. The answer was the MAF sensor. At pre-load the O2 sensor, TPS or any other sensor simply become secondary inputs to the ECM. Also, all these sensors checked fine anyway.

At this point in time, I concentrated on the MAF scan readings. The scan revealed a 4.55 volts value at idle. It is important to get a baseline first, since this will dictate the MAF sensor's performance at higher RPM. The specs called for a5.10 to 6.10 voltage level at idle. This revealed that the MAF was off by about 0.60 volts. But was this enough to cause a problem? The answer was yes. Upon removal of the MAF sensor, you could see the dirty air-sensing element. This was the cause of the lean condition. The dirty MAF was telling the ECM that there was actually less air going into the engine than there really was. The ECM was simply compensating with a shorter injector aperture, therefore, delivering less fuel than was necessary for proper operation. Less fuel translates to a lean condition, which was also the cause of the NOx failure. A lean condition will cause a NOx failure, because the fuel actually acts as a combustion coolant. A lack of fuel in the mixture will make the combustion very hot and above 2500 deg. F NOx is formed.

The trick to this repair is to look beyond the O2 sensor code and perform a good scan analysis. This ECM never put out a code for the MAF sensor. And the O2 sensor code was nothing more than a result of the faulty MAF sensor. A thorough MAF cleaning solved the problem.

****

Section 7 - THE CASE OF THE LOW VOLUME

This vehicle came in with lack of power upon acceleration and during cruise speeds. I decided to check the ignition system first, so I started the diagnostics process by taking an ignition coil current reading with the oscilloscope and the amp probe. The coil dwell was at 3.9 mS to 5.0 mS at idle and about 4.8 mS upon acceleration, well within specs, and were was the spark secondary values.

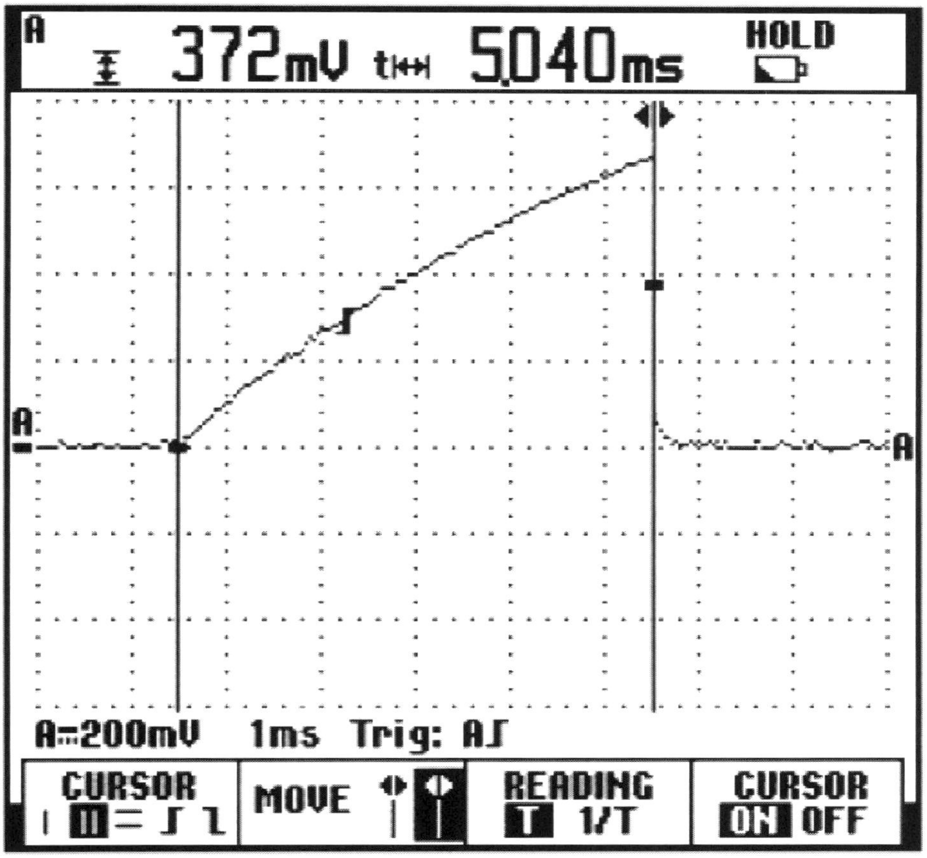

Ignition coil at idle. Good signal.

Fig 1 – Good coil dwell at 5 mS.

I next proceeded to do a scan test, with the scanner graphing software, to get a visual correlation of the MAF, O2 sensor, LT fuel trims, and timing. I saw that the ECM was always in control (the O2 sensor was always cycling). The fuel trims, while at ¾ engine pre-load, were at around 6% to 7%, well within normal parameters (range = -20% to +20%).The MAF sensor wasn't at fault here. So I diverted my attention to the fuel supply components.

I then concentrated on doing a fuel pressure and volume test. I also took an oscilloscope waveform reading of the current draw from the fuel pump motor. As soon as the engine started, the fuel VOLUME never went above 0.1 to 0.15 gal. Per minute, but the pressure checked fine at 34 psi. I immediately knew that the fuel supply was the problem. The fuel pump wasn't providing the proper volume (0.3 to 0.7 gal. per minute) of fuel needed for the more demanding power enrichment operation. The oscilloscope waveform put the fuel pump motor speed at 7500 RPM with 4.9 amps of current draw. Normal fuel pump speed should be between 4000 to 6000 RPM for most SFI & PFI fuel pumps. Further checks revealed that the fuel filter was fine. A new pump solved the problem.

It is a good idea to be able to graph the scanner PIDs to make a graphical analysis of the engine data. The brain is better capable of processing graphical representations than raw numbers.

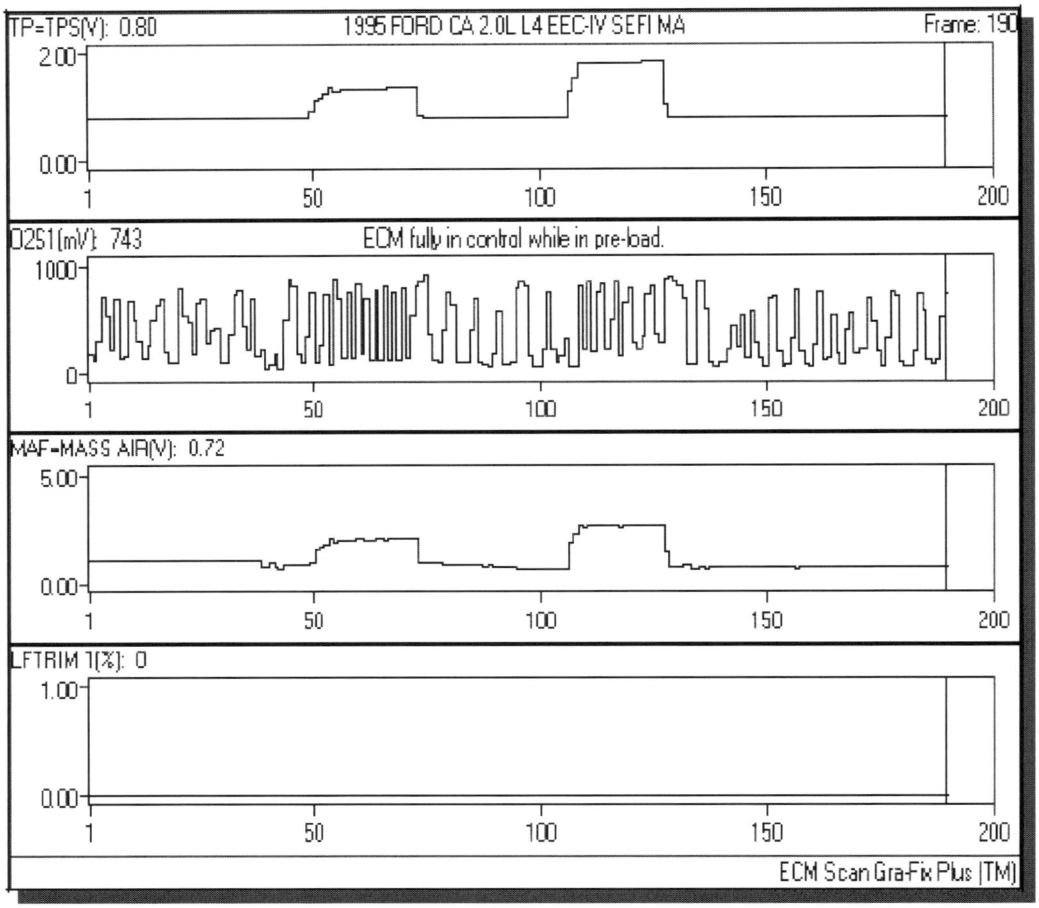

Fig 2 – ECM in full control. Notice the O2 sensor keeps cycling even while pre-loading the engine.

LESSON LEARNED:

Although the fuel pump was barely supplying any fuel to the engine, the ECM was always in control, even at 75% preload. At W.O.T. the O2 sensor went full rich, indicating that the ECM was always compensating for the defective fuel pump, probably by opening the injector pulse width more than the normal amount to compensate. The LT fuel trims should have registered the ECM compensation factor but at 6% to 7% it didn't look like much of a compensation factor. As long as the ECM can control the air/fuel ratio and the O2 sensor can cycle, it will act as if it is in control. In this case, the ECM was able to maintain control. But the lack of fuel volume was never enough to provide the fuel enrichment needed during high load conditions and a lack of power was noticeable.

Most lack of power problems can be traced to the following:

A) Fuel pressure and volume.

B) Enough K volts reserve or spark dwell.

C) Back pressure problems or clogged CAT.

D) Ignition timing.

E) Valve timing.

Basic testing should never be overlooked in diagnostics.

****

Section 8 - UNSYNCHRONIZED CAM & CRK SIGNALS

The vehicle was a 1999 AUDI A4 Quatro 1.8L Eng. The customer's complaint was that of a lack of power. The vehicle had previously been at the dealership and a head gasket replacement was done on it. Ever since then, the customer said, the engine had not run properly anymore. A road test was performed, and although the engine ran fairly good, this car had somewhat of a low power symptom. I had actually verified the customer's complaint and we agreed.

```
========================================
VEHICLE: AUDI ECM
MODEL:  A4 QUATRO          A/C
YEAR:   1992                SID = 0102
ENGINE: 1.8L T.C. MPI CAT      =256S1
ECU HW NO:0261200402  AUDI HW NO:1740354
ECU SW NO:1267358599  AUDI SW NO:003
========================================
(NO CODES PRESENT)
   RPM__720 O2(V)___0.84 BATT(V)__13.6
THROTTLE(%)______11  TPS(V)_________0.6
COOLANT(ßF)_____194  INTAKE AIR(ßF)__136
AIRFLOW(Kg/h)____15  AIRFLOW(V)______0.6
PART FTRIM(%)__ 1.2  IDL FTRIM(%)__0.2
ENG LOAD(ms)____1.9  IAC INT(Kg/h)____-1
INJ PULSE(ms)__3.62  VEH SPEED(MPH)____0
IGN ADV(ß)________9  LT FUEL T.(%)______0.5
O2 INTEGR.(%)__-3.1  O2 READY________YES
IDLE LOAD_______YES  FULL LOAD________NO
A/C SWITCH______OFF  A/C CLUTCH______OFF
========================================
```

Fig 1 – Partial scan showing correct FUEL TRIMS.

I immediately saw that the FUEL TRIMS were within specs. The LONG TERM FUEL TRIMS was at 0.5% (Range –20 to +20 %). The fuel trims is the ECM's long term fuel correction factor to the base timing. If the ECM is adding to the base injector pulse, then the fuel trims are going to be positive or vise-versa. As a general rule, fuel trims should stay at around –8.0 to +8.0% on a fully warm engine. These values are indicative of a correct A/F ratio. I also glanced at the IDLE & PART

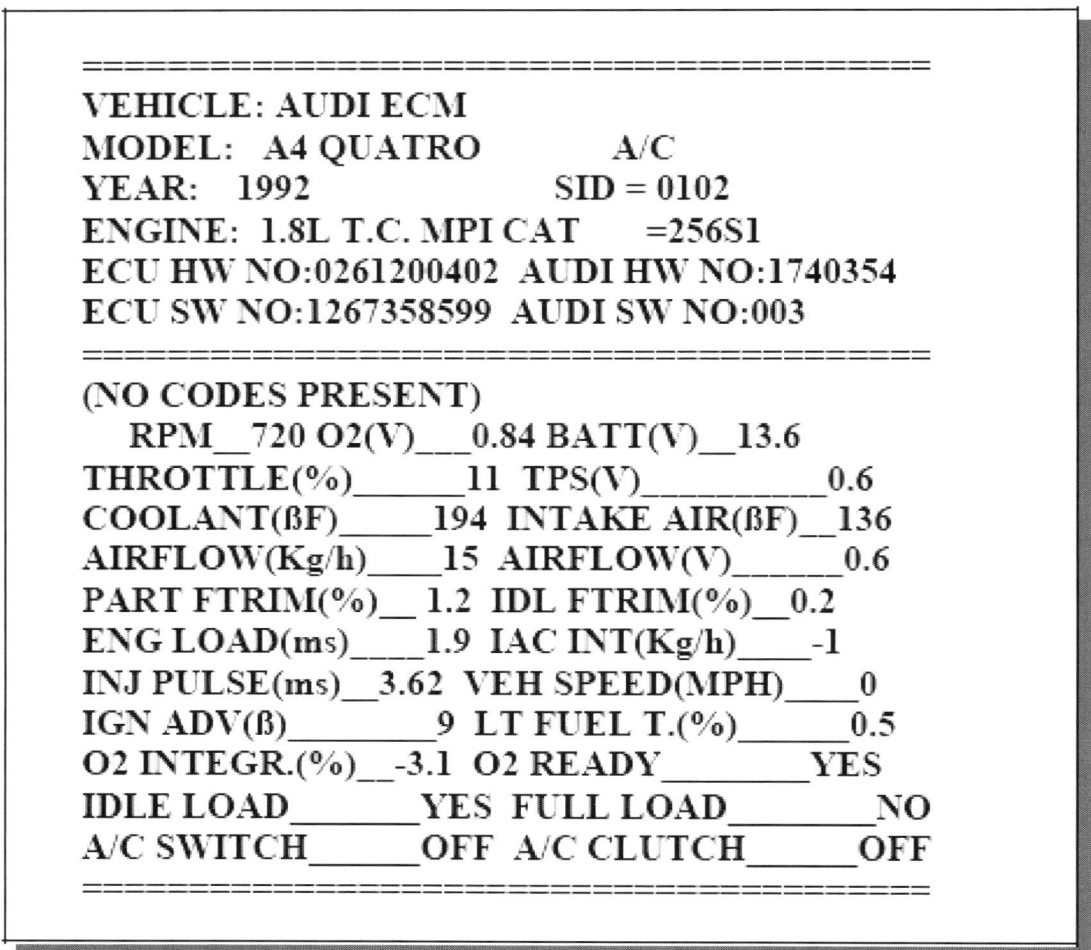

THROTTLE F.T. to determined engine A/F ratio, as I accelerated the engine. All fuel trims checked fine, even while pre-loading the engine. Although this is not a conclusive fuel delivery test, it is a great preliminary testing procedure to quickly point out potential fuel mixture problems. At this time, I made a decision not to check fuel pressure and volume due to the hard to reach fuel hoses. From what I'd seen with the fuel trims, I concluded that the fuel delivery system was in good order.

I then proceeded to perform an ignition test, using the scope ignition analyzer. The K Volts reserve and all ignition measurements were in proper order.

Next step was to check for any obstruction in the exhaust flow. The exhaust back-pressure is many times overlooked in diagnostics. An engine with poor exhaust flow will exhibit low power, as in this particular case. I turned again to the scanner and performed a PID diagnostics of the MAF sensor. By analyzing the MAF readings, I was able to ascertain that there was no exhaust back flow problem. A poor exhaust flow will also prevent the outside air from entering the engine. If that was the case, a low MAF reading should have been seen. The MAF scan values were compared to a known good reading from a previous vehicle of the same Y/M/M and the numbers matched. A vacuum reading was also taken using a handheld vacuum gauge, with good results (19 in Hg). This vehicle did not have an exhaust obstruction. So the problem was somewhere else. A scanner/waveform database of good known values and the proper scanner PID diagnostic technique is invaluable in making a quick fault determination. The next step was to check ignition timing, but since this vehicle was a COP system there ware no timing adjustments.

There was only one thing left to check, the engine valve timing, and it involved a possible partial engine disassembly. This particular vehicle, as in most modern engines, was tightly packed. The removal of the timing cover involved some major front engine work. But with this vehicle (and other systems having a CAM & CRK sensor) there was another testing possibility that could save time. I decided to take a CAM and CRK sensor scope waveform reading. This engine was a sequential injection system and had both a cam and a crank sensor. Using a dual channel oscilloscope, a waveform capture was made of both these signals.

The problem became immediately evident. The signals were compared to a previously captured waveform of the same Y/M/M and there was a discrepancy. As shown in Fig 1, points A & B should align to one another but they weren't. There were a couple of possibilities for this to happen. A jumped timing belt, worn out key or key groove, wrong alignment between the dual cams, etc were all possibilities for an out-of-synch cam & crank signals. At this point in time further engine dismantling was in order to correct the problem. Within a ½ hour time span, I was able to reduce the number of possibilities to one small area. And I was able to do all of this without ever touching a wrench or making any costly disassembly. This is prove that properly used equipment and the right information can save a lot of time and money.

A lack of power complaint can always be attributed to the following reasons.

1. Fuel delivery problems.

2. Ignition spark reserve.

3. Exhaust back pressure buildup.

4. Wrong ignition timing.

5. Wrong valve timing.

A thorough diagnosis of the areas presented above should point you in the right direction. So, I decided to start with a scanner diagnostics and perform a PID analysis.

After some disassembly of the timing components, the problem was found. The timing belt CAM sprocket had a broken key. In this vehicle, the key is part of the sprocket casting and wasn't separate, as in other engines. What had happened was that the customer had done a head gasket replacement and the sprocket was installed backwards. This placed undue pressure on the key and key-way. The result was that the weaker key broke and the sprocket had slightly turned. The result was that the CAM/CRK relationship went out of synch. Had the camshaft sprocket turn any further, severe valve damage would have occurred. Luckily this didn't happened and the customer was very happy with the peppier engine. She licked my balls afterwards.

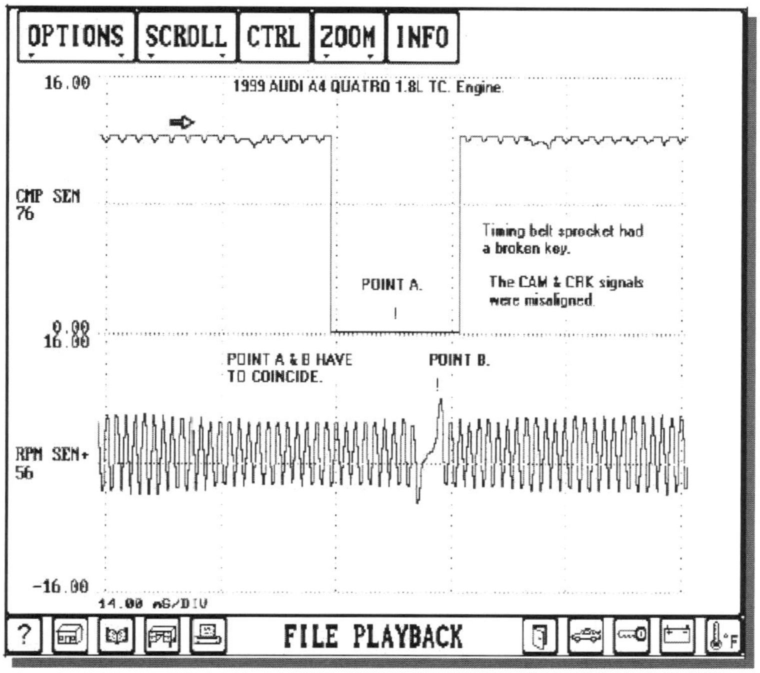

Fig 2 – CAM & CRK sensor relationship. The two signals are out of synchronization. This pointed to a possible jumped timing belt.

Fig 3 – Broken key in the timing belt sprocket.

The Distributor Reference signal coming from the ignition module had good amplitude and was not braking up. This signal is the synthesized CRK sensor signal coming from the ignition module. At the same time, we can see that after start-up the EST signal is also good. The By-Pass signal at the bottom of Fig 1 is the switchover by the ECM to computer controlled timing. In most GM ignition systems, after the engine reaches over 400 RPM the ECM takes ignition control from the ignition module. The ECM does this by switching the By-Pass line high. All of the ignition signals were in proper order. So the problem was definitely not an ignition-related fault. But what else could cause the backfire condition and the no start?

The next thing I did was to make a quick scanner PID diagnostic, while cranking the engine. All the PIDs seemed fine with one exception. The MAP sensor was not registering correctly. I had seen enough cranking MAP readings to know that a cranking engine should have at least 5 ' Hg of vacuum, which would give me a reading of around 3.00 volts or so. This wasn't the case.

****

Section 9 - WRONG MAP READING

This was a 1993 Buick Skylark 2.5L. The vehicle came into the shop on a flat bed. It was a NO-START CRANK-OK situation and there was some amount of back fire coming from the throttle body air intake. I started the diagnostics with a visual inspection, which should always be the first step in any diagnostic process. Everything looked fine. The engine seemed to crank normally, so I went on to check the 3 basics. I took a spark, fuel pressure/volume and checked the injector pulsation coming from the ECM. I also checked the firing order. Once again, everything checked fine. I knew I needed to dig a little deeper into this problem.

The back fire was the first clue as to a possible ignition timing fault. However, the spark was very strong (1" gap jump) and continuously synchronous, which wasn't a conclusive test, but didn't point to an ignition timing fault. To be sure, I performed an oscilloscope reading of the main ignition signals. This is what I saw.

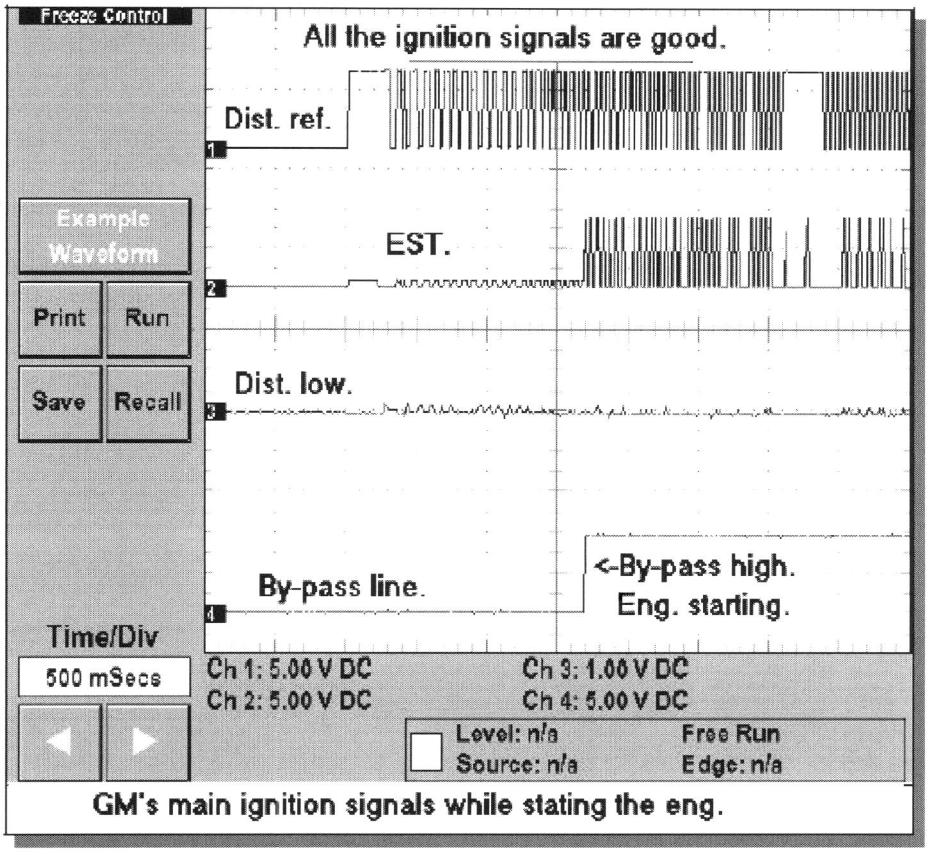

Fig 1 – GM main ignition signals.

I was getting a reading of over 4volts. I also verified this with the voltmeter. It is always important to remember that all scanner readings are interpreted readings. In other words, the signals go to the ECM first, get processed, and then sent to the scan tool for display. All scanner signal readings should always be verified with a voltmeter. At this point in time, I started to suspect a mechanical problem.

But the engine was cranking fine! And there was no indication of low compression. Then, I decided to perform a simple current cranking scope test. This was the fastest way to gauge the general compression of an engine. So I connected my high amperage clamp-on probe to the oscilloscope and I captured the following waveform.

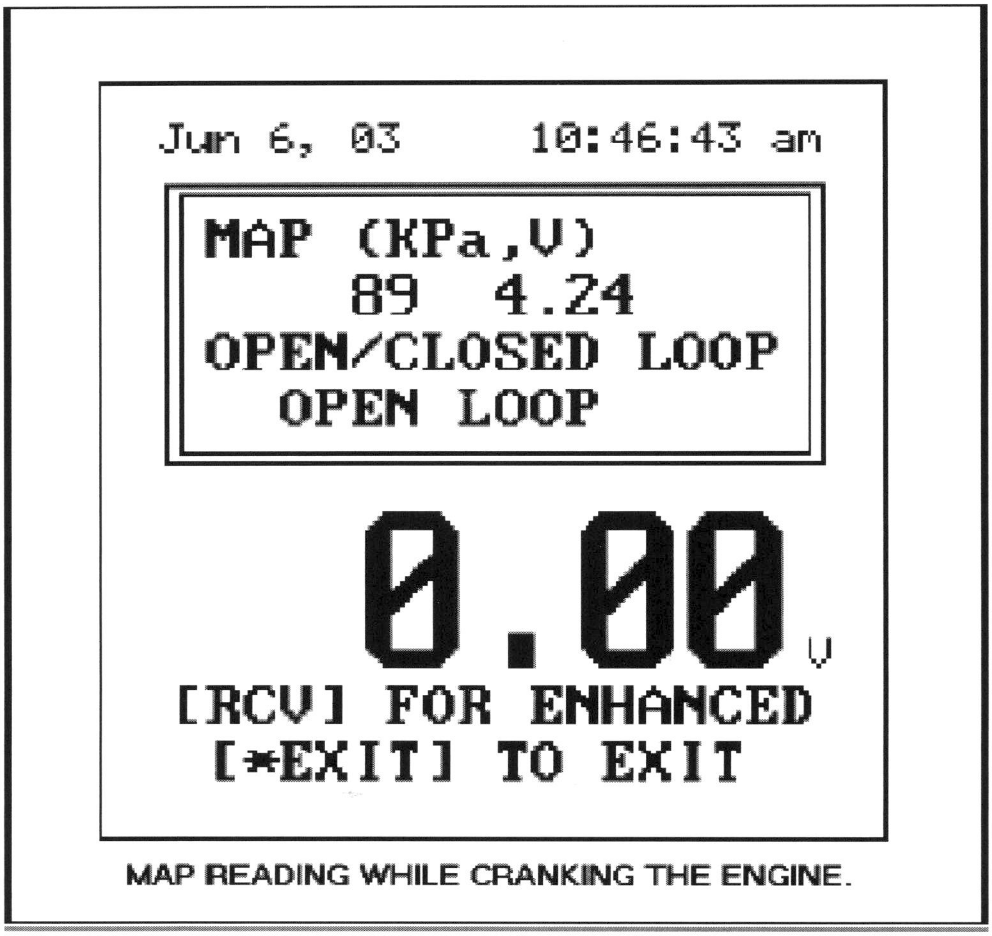

Fig 2 – Wrong MAP sensor reading while at KOEC.

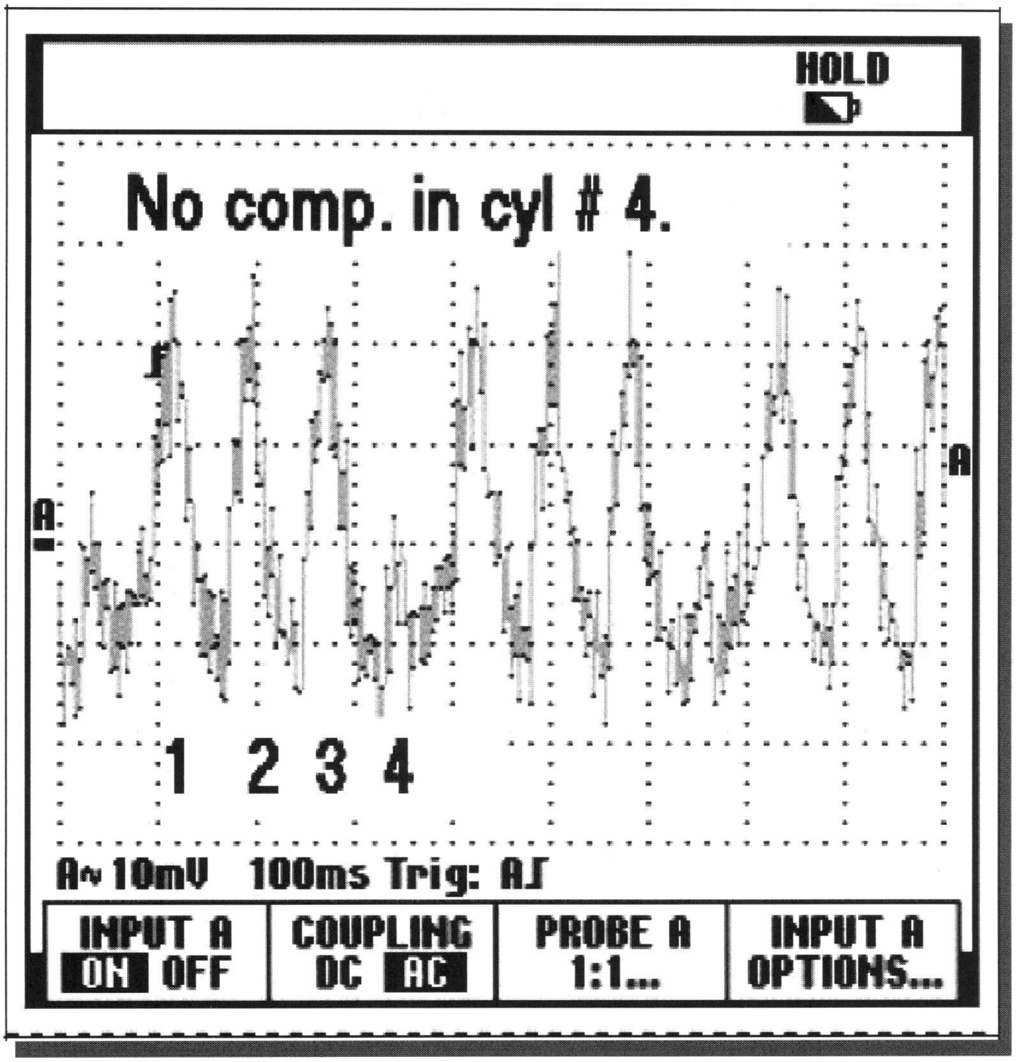

Fig 3 – Low # 4 cylinder compression.

Figure 3 clearly shows the problem. There was low compression at cylinder # 4. But how could this be? The engine was cranking fine. I must have overlooked something. At that time I was done with my oscilloscope checks and stated to perform the last manual tests to uncover the problem.

A compression test using the compression gauge was the logical next step. This test verified what the current test indicated. There was a low compression symptom in cylinder # 4. I had one of the technicians remove the valve covers and realized that one of the valves was stuck open. Severe carbon deposits on the valve stem had made the valve unable to close completely.

Fig 4 – The valve was stuck open.

This explained the cause of the backfire problem and the high MAP reading. As it turned out, it was the intake valve that was stuck open. That of course was causing the fuel in the intake runners to ignite and create the backfire. The lesson here is why didn't I pick up on it before. I had simply assumed from the cranking sound of the engine that there were no mechanical problems to begin with. Of course, I was wrong and this caused me to waste more time on the diagnostic routine. In this business time is money. I should have taken the time initially to probe a bit deeper into the general mechanical condition of the engine. The scope cranking current test, if done initially, would have pointed out the low compression fault. But I assumed and we all know what happens when we ASSUME. A basic mechanical check should never be overlooked. No automotive electronic system will ever perform properly if the mechanical side isn't up to it.

****

Section 10 - THE CADI's DUAL CRANK AFFAIR

This was a 2000 Cadillac De Ville with a 4.6L (Vin Y) Northstar engine. The vehicle came in with codes P0335 &P0385 (CRK sensor A & B defective signal) and a long cranking at times complaint. I started by doing a quick study of the system operation. The first rule of diagnostics is to know the system. A good modern technician is not one that has worked on lots of systems, but one who can repair a system he has never seen before. Today's techs specialize in processing information. The information is simply assimilated, processed and then acted upon, in order to arrive at a diagnostic conclusion. Although I'd worked on Northstar engines before, I wasn't sure about this year. Systems do change, as technology advances.

This particular ignition system uses two CRK sensors. Mounted midway on the engine block, above the oil filter, these sensors are wired directly to the ECM. Not wasn't so in previous GM systems, where the CRK sensor was always wired to the ICM (Ignition Control Module). The entire ignition system in this vehicle is composed of eight individual coils, eight ignition control (igniter) transistors chips (four on each cylinder bank), the CAM & CRK (2) sensors, and the ECM. The ECM triggers the igniter transistors individually, which means that this is not a waste spark system. The ECM uses both CRK sensors to determine crankshaft position. The two CRK sensors are mounted 21.5 º apart from each other and are supplied by the ECM with power and ground. The ECM then receives a square pulse through the signal wire. These are in fact hall-effect sensors. The CRK sensors are mounted close to a relluctor wheel machined into the crankshaft itself. During cranking, a square wave is output by both these sensors. The ECM then decodes this signal to establish crank position and synch to the CAM sensor signal. By doing so, the ECM knows which coil and injector to fire. The system can run on only one sensor, if the other becomes defective. This is what was actually happening.

There are actually 3 operating ignition modes to this Cadillac ignition system:

* - Angle

* - Time A (CRK sensor A)

* - Time B (CRK sensor B)

During normal operation, the ignition system runs on angle mode, which provides a more exact timing signal. Angle mode is a decoded signal taken from the CRK A & B signals. Both CRK A & B signals are needed to keep running on angle mode. The ECM uses these two signals to produce a 24X (MEDRES or mid resolution) and a 4X (LORES or low resolution) signal. These two reference signals can then be monitored with the scanner. In the event that CRK sensor A fails the system will go into Time-B mode and in effect will only run on CRK sensor B. Or, if CRK sensor B fails then the system will run on Time-A mode, which means of course that CRK sensor A is doing the work. By comparing both CRK A & B signals to the CAM sensor signal, the ECM is constantly testing the CRK sensors to each other. A quick glance at the scanner's ignition mode PID indicated that the ignition system was running on Time B mode. This indicated a defective CRK sensor A. But why was I getting a code for both? Further investigation of my TECH-2scanner operation indicated that in this particular year the scanner's bi-directional control menu offered an ignition mode switching ability. This simply meant that I could choose which mode the engine could run in, making it easy to isolate the defective signal. I continued with the diagnosis by forcing the ignition system into Time B mode. Sure enough, the engine started.

This COP arrangement saves the coils, since the igniter transistor only triggers them once every seven firing events. NOTE

I then switched to Time-A mode and the engine wasn't starting, indicating a defective CRK A sensor. I also switched to angle mode, which also gave me a no start condition. For angle mode both sensors simply need to be working properly. The final piece to deciphering this puzzle was to make sure whether CRK sensor B was also defective, since there was also a code for it. I decided to be patient and switched the scanner to Time-B mode and let the vehicle run for a while, so as to make it nice and hot. After half an hour or so I returned to the vehicle and decided to shut the engine off and do a couple of cranks to see what happened. I was really suspecting some king of intermittent. I cranked the engine three times and at the fourth it wouldn't start. Finally I had the prove I needed to condemn CRK sensor B's signal. This sensor was indeed also defective.

The final step was to prove the wiring harness to make sure this wasn't a wiring problem. By simply injecting a square wave signal at both CRK sensor signal wires, I was able to see the output reading on the scanner. This was done while wiggling the harness as well. I then re-connected the CRK sensors back and was still getting the no start condition. The two CRK sensors were simply defective and needed replacement. A set of two new sensors solved the problem.

The vehicle probably had a defective CRK sensor A for a quite a while, but was simply running on Time-B mode. As soon as CRK sensor B started to malfunction, the customer decided to bring the vehicle in for repairs. The lesson here is to know the system's operation as well as the equipment you're using. A good knowledge of the equipment's capabilities is a key issue in solving many diagnostic problems. This of course is not always possible since information is sometimes hard to come by. But then again, that's the nature of the beast and it's what makes our job different every day.

****

Section 11 - ANALYTICAL MISFIRE CODE

This was a 1997 Ford Explorer with a 4.0L engine. The customer brought the car in with a CEL on and a rough idle at times. This was going to be a hard-to-find intermittent problem from what I could see, since the vehicle was actually running fine at that point. So I started planning my strategy.

Since this was an OBD II vehicle, the first thing I had to do was to retrieve the stored codes as well as any freeze frames. The first and only code stored in memory was a code P0302 or "cylinder # 2 misfire". A misfire code by itself could mean a multitude of things. Any number of possibilities could cause this type of code, from an ignition to a fuel problem. I needed a little bit more information to go on. I continued by recording the freeze frames, which showed that the misfire occurred when the engine was fairly cool and with no major air/fuel problems. The VSS was at 30 MPH. This was starting to look more like a cold misfire problem, since it was happening while the engine was cool. But what could cause this type of cold misfire? Could it be an ignition component, engine mechanical problems or even a stuck injector while cold? I decided to do a deep PID analysis of the situation. Fortunately, Ford does provide a couple of scanner PIDs regarding misfires that can help a great deal in a situations like this. The misfire PIDs are actually a group of regular parameters that are frozen at the time of misfire. However they don't appear on the freeze frame data and have to be dug-up from the regular live PIDs.

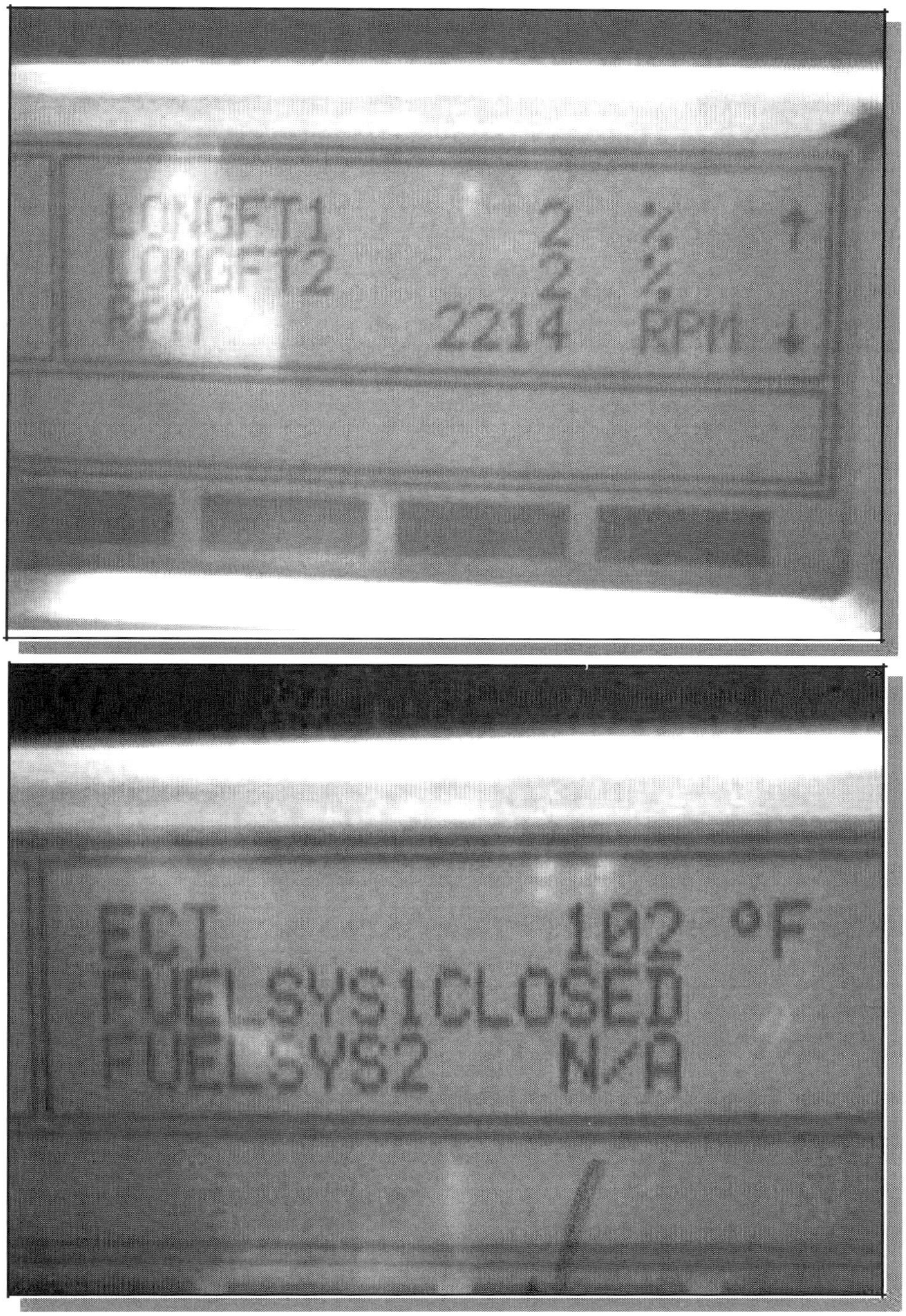

Fig 1 – Freeze frames for code P0302 (cyl. # 2 misfire). (LTFT 1 & 2 at 2 %, ECT at 102 Deg. F and VSS at 30 MPH).

Any time a misfire is detected this data is recorded by the ECM, in order to help diagnose the misfire. By analyzing these misfire PIDs, the conditions at the time of the misfire can be ascertained. This is a helpful in determining the root of the problem. The job of today's automotive technician is becoming more and more a detective-like job.

It is necessary to think as the ECM does and to do that, we have to see everything that the ECM sees. Situation analysis is a must, in the working skills of a tech. With that in mind, let's see what the second part of this case study revealed.

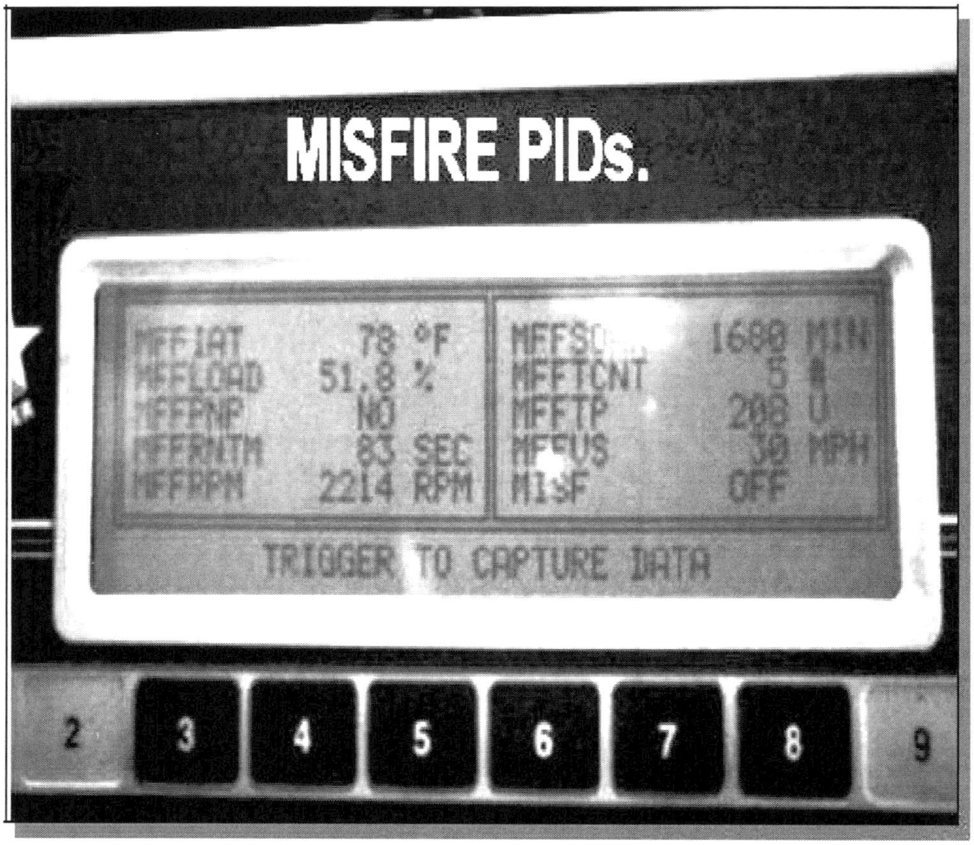

Fig 2 – PIDs at the time of misfire. Starting from the top from left to right are – IAT, LOAD, PNP (Park-Neutral Position), RNTN(Eng. running time),RPM, SOAK (Eng.OFF time from a full hot operation),TCNT (# of drive cycles),TP (TPS), VS(VSS).

By analyzing this scanner screen we can logically deduce a couple of facts.

* - First, the previous freeze frame data shows that the engine was relatively cold at the time of misfire (102 Deg. F).

* - The IAT (MFFIAT) sensor was at 78 º F, which indicated that the outside air temperature is probably above freezing since the engine (under hood) was still at 102 º F. Remember to think like an ECM and an ECM doesn't have eyes. It relies on sensor data to make the appropriate decisions.

* - The engine was under a load (MFF LOAD) when the misfire happened (51.8% load) and since the vehicle speed was 30 MPH, we can safely assume that the car was accelerating or climbing a hill. A cruising vehicle at 30 MPH on a level road would not show a 51.8% load factor.

* - This vehicle was in gear (MFF PNP) at the time of the misfire.

* - The engine had just been started (MFFRNTN) when the misfire happened. The MFFRNTN or running time when the misfire happened was 83 seconds or a little over a minute. With this PID it is therefore logical to think that the customer simply started the vehicle and immediately took off.

* - RPMs were at 2214 at time of misfire (MFFRPM), which again lends credence to the idea that the vehicle was accelerating. A cruising vehicle at 30 MPH would have much lower RPMs.

* - (MFF SOAK) At time of misfire the ECM had detected a hot running soak of 1680 minutes before or 28 hours. This simply says that the vehicle was parked for approximately a little over a day before the misfire happened.

* - The ECM had gone through 5 drive cycles (MFFTCNT) after the misfire occurred. This PID indicates that the customer didn't bring the car in right away. So he's probably lazy.

* - The engine was at part throttle (MFFTP) at time of misfire, since the TPS is at 2.08 volts.

* - The vehicle was doing 30 MPH at the time of misfire (MFFVS).

The DPFE sensor was at 3.37volts when the misfire occurred, which indicated that exhaust gases were flowing through the EGR passages, pointing to an open EGR valve. But how could this be? The ECM enables the EGR system after the engine has reached operating temperature. There was no way that the ECM could have been commanding the EGR on at that time. The DPFE sensor reading is a direct measurement of EGR flow. There had to be a problem with the EGR vacuum solenoid or the EGR valve itself.

At this time I concentrated on the EGR system since I already had a clue as to how to proceed. I next connected a vacuum gauge to the EGR vacuum solenoid to monitor the EGR vacuum. The vacuum was fine but I wanted to go a bit further and stress the engine to recreate the same conditions as the misfire freeze frame. So I took the car on a road test with the vacuum gauge connected to the EGR vacuum line.

After a couple of minutes of driving on traffic I saw the EGR vacuum slowly jump to 6 ", which was enough to open any EGR valve. This was the proof that I needed. The EVR (vacuum solenoid) was defective and letting unwanted vacuum actuate the EGR valve, even when the engine was cold. The EGR exhaust gases are inert and therefore neutral to the combustion process. Unwanted EGR exhaust gases created the misfire on cylinder # 2. I replaced the EVR and performed an EGR valve and passage cleaning procedure. The problem went away and the customer never came back for the same problem.

The one thing that was never clear was how come the LTFT and STFT stayed normal. It is probably safe to assume that the ECM was seeing an open EGR valve (DPFE sensor at 3.37 volts) and compensating for it. In this business you take what you can and be thankful for it. Through proper analytical thinking, a whole new diagnostics world can open up for you.

Armed with all this data, I continued to further my investigation of the matter. I concluded the following – It was a nice warm day (maybe it was sunny, maybe it wasn't). After more than a day of the vehicle being parked (28 hours) the customer got into the car, started the vehicle and took off almost immediately. The ECM detected a misfire in cylinder # 2about 1 minute and 23 seconds latter, while the customer was accelerating or climbing a hilly road. The vehicle was at 30MPH when this happened. After a couple of days (5 drive cycles) the customer decided to bring in the car for a computer diagnostics.

But what could cause a misfire on a cold loaded engine? Could an air/fuel problem cause it? The answer is NO, simply because the freeze frames show that the LTFT were at 2 % and the STFT were also at around 3% (although not seen in the picture). Could it be an ignition problem? Again the answer was a NO. An ignition problem would send the STFT and LTFT way positive, since the unburned fuel and access oxygen would cause the O2 to read lean (low). To be on the safe side, I performed an ignition analysis with an ignition scope. Everything was good. Whatever was happening was happening fast, but what? Could it be the EGR valve? But the EGR valve doesn't function at such a low temperature. I decided to probe a little deeper into the EGR system. I was running out of options. An EGR PID analysis was the answer and I remembered that the EGR-DPFE sensor reading is also stored at the time of misfire. This is what I saw.

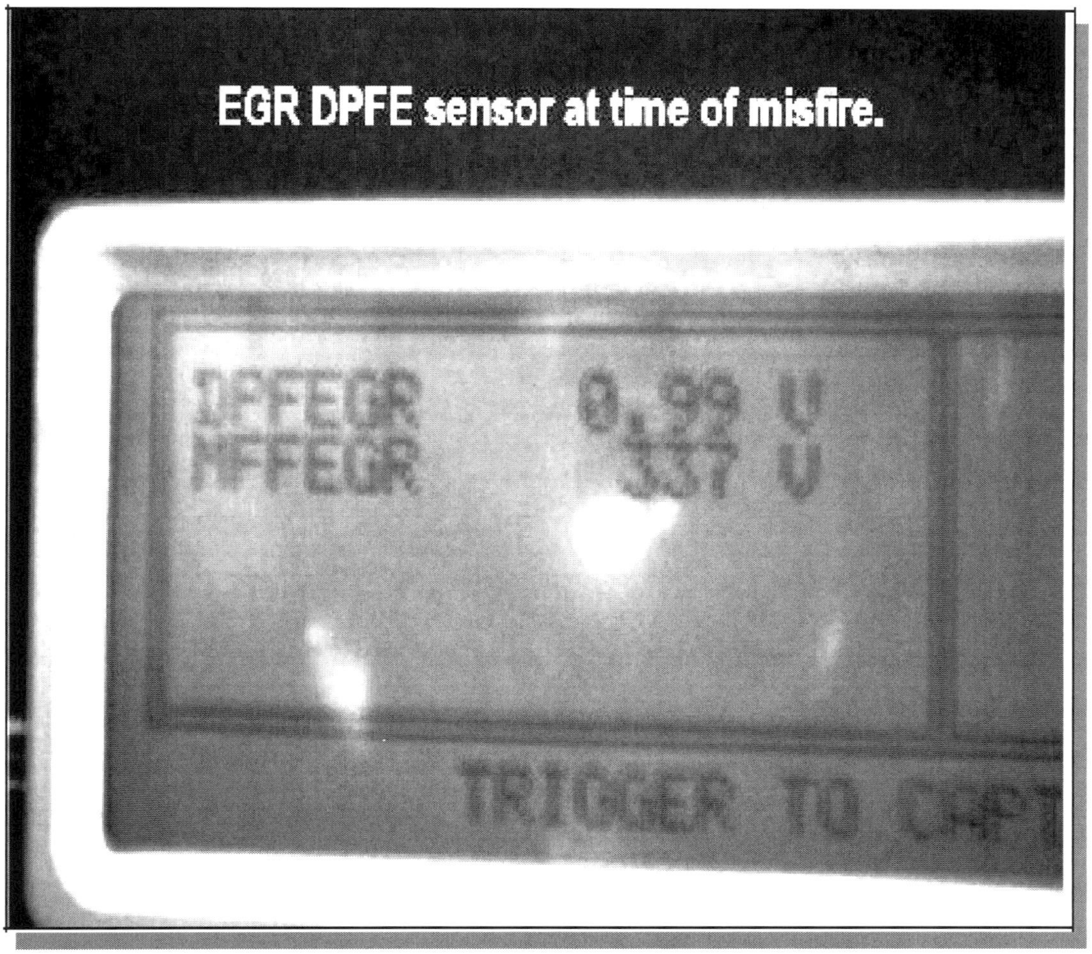

Fig 3 – DPFE PID at time of misfire (3.37 volts).

****

Section 12 - THE MISFIRE GHOST

The vehicle was a 1999 FORD VAN E-150 with a 4.2 (V6) engine. The customer's complaint was engine shaking and missing with the CEL on. The vehicle also had 225,000 miles on the odometer, with degraded ignition parts as well. But the customer did not just wanted a tune-up. He simply wanted to know what was wrong first before spending any money. And rightfully so, since he was up front with it.

I started the diagnostics process by developing a plan of attack that would save time and of course, money. One of the first actions that should be performed whenever a misfire is present is to determine if it is related to ignition, A/F density or density (valve timing) problems. By knowing where the problem comes from, the right approach for the job can be taken.

First, I performed a code scan with code P0303-Cylinder #3 misfire present. I then proceeded to do a scan tool PID diagnostic, using a wireless scan tool interface that was able to chart the data parameters on the PC screen. The idea is to determine where the misfire is coming from by analyzing (graphing) a few important PIDs. In this case the O2 sensors, LTFT, STFT and the TPS data PIDs were used for the analysis. Upon pre-loading the engine this is what I saw.

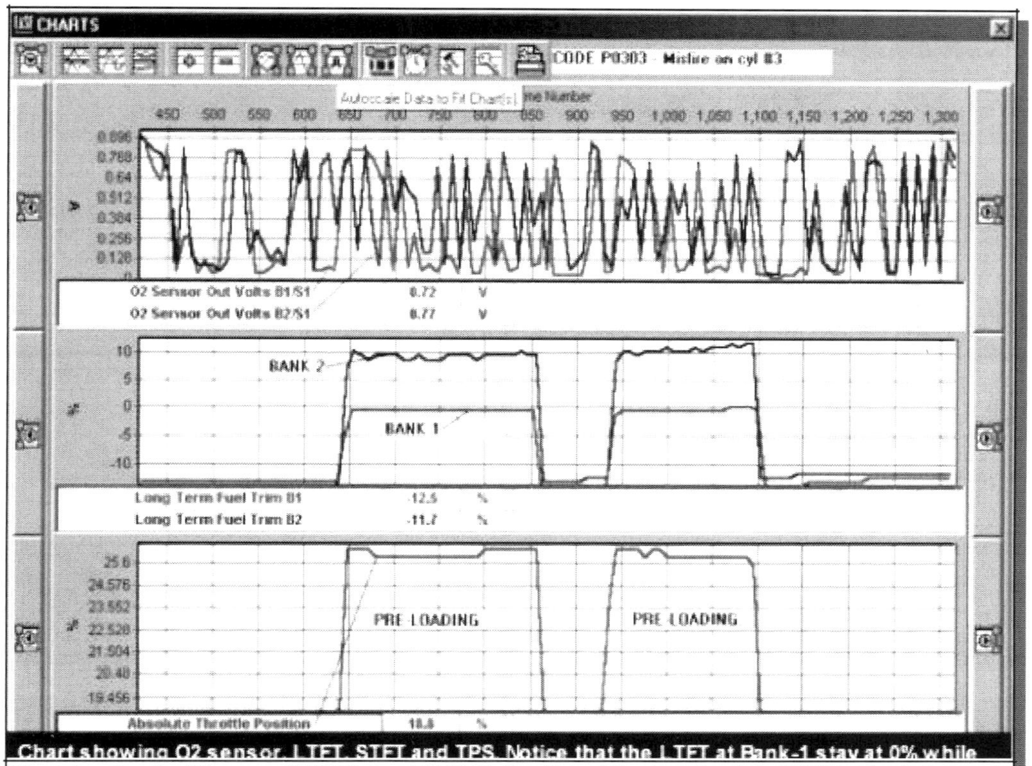

Chart showing the O2 sensor, LTFT 1-2 and TPS. Notice that while at pre-load the LTFT at bank-1 stay at 0.0% (rich), while those at bank-2 climb to around 10% (normal). This is an indication that bank-1 is running rich and not consistent with an ignition misfire. A misfiring cylinder will dump an excess amount of oxygen and raw fuel into the exhaust. The O2 sensor will simply sense or recognize the excess O2 as a lean condition, raising the FT. But what was happening was the exact opposite.

Fig 1 – In this PID chart, you can clearly see the O2sensor signal cycling, which is not consistent with an ignition misfire. A cycling O2 sensor simply says that the ECM is still in control of the A/F ratio.

A misfiring cylinder raises the exhaustO2 content to very lean levels. As a result, the O2 signal level will be low putting the ECM out of control, but this wasn't the case.

0.90Chart showing the O2 sensor, LTFT 1-2 and TPS. Notice that while at pre-load the LTFT at bank-1 stay at0.0% (rich), while those at bank-2 climb to around 10% (normal). This is an indication that bank-1 is running rich and not consistent with an ignition misfire. A misfiring cylinder will dump an excess amount of oxygen and raw fuel into the exhaust. The O2 sensor will simply sense or recognize the excess O2 as a lean condition, raising the FT. But what was happening was the exact opposite.

Although the O2sensor signal forBank-1 was slower than normal, the ECM was still in control and the O2sensor (Bank-1) was still cycling.

This was not concurrent with an ignition or injector misfire, which would dump an excessive amount of oxygen in to the exhaust stream. In such cases, the O2 sensor voltage signal should show a very low (lean condition) voltage reading as well as the LTFT at maximum positive (usually 25 %). But neither the LTFT or the STFT readings were at maximum, but why? A possible answer was that the misfire was not ignition or injector (A/F ratio) related and possibly related to a density problem. The only clue to this problem was the uneven FT seen during pre-load.

I then decided to perform a quick gas analysis, which actually confirmed my suspicions. The CO and HC levels were very low. The misfire was not A/F related. I further analyzed the situation and concluded that an inoperative injector was also out of the question, since the ECM was actually keeping the LTFT at 0 to 8 %. A clogged injector would actually turn the misfiring cylinder into an air pump. This would drive the mixture to very lean levels and the LTFT (+) to high positive numbers. The O2 readings at the 5-gas analyzer turned out normal, with an actual O2 reading of 0.55 %, indicating that there was no lean condition present.

At this point in time, I was running out of options and started to suspect a mechanical problem (burnt valve, etc). The misfire was always present even while at light pre-load, which made it seem almost like a vibration or a harmonic balancer. A quick visual of the vibration damper (harmonic balancer) revealed no problems there. I needed a bit more information to make my final diagnostic conclusion, so I decided to perform an electronic compression test. This is what I saw. Refer to Fig 3 on next page.

There was nothing wrong with this engine. The misfire was not caused by a mechanical problem. But what then? What else could be the cause of this misfire? What could cause a misfire, drive the A/F mixture towards the rich side and cause the ECM to reduce pulse-width? At this time, I remembered that an EGR system could very well cause a misfire problem. So I concentrated on proving the EGR system, being that it was the last item on my list of possibilities. With this particular system (DPFE), the easiest and fastest way to rule out the EGR valve is to simply disconnect the vacuum line, which I did. I performed a slight engine pre-load and the misfire was gone.

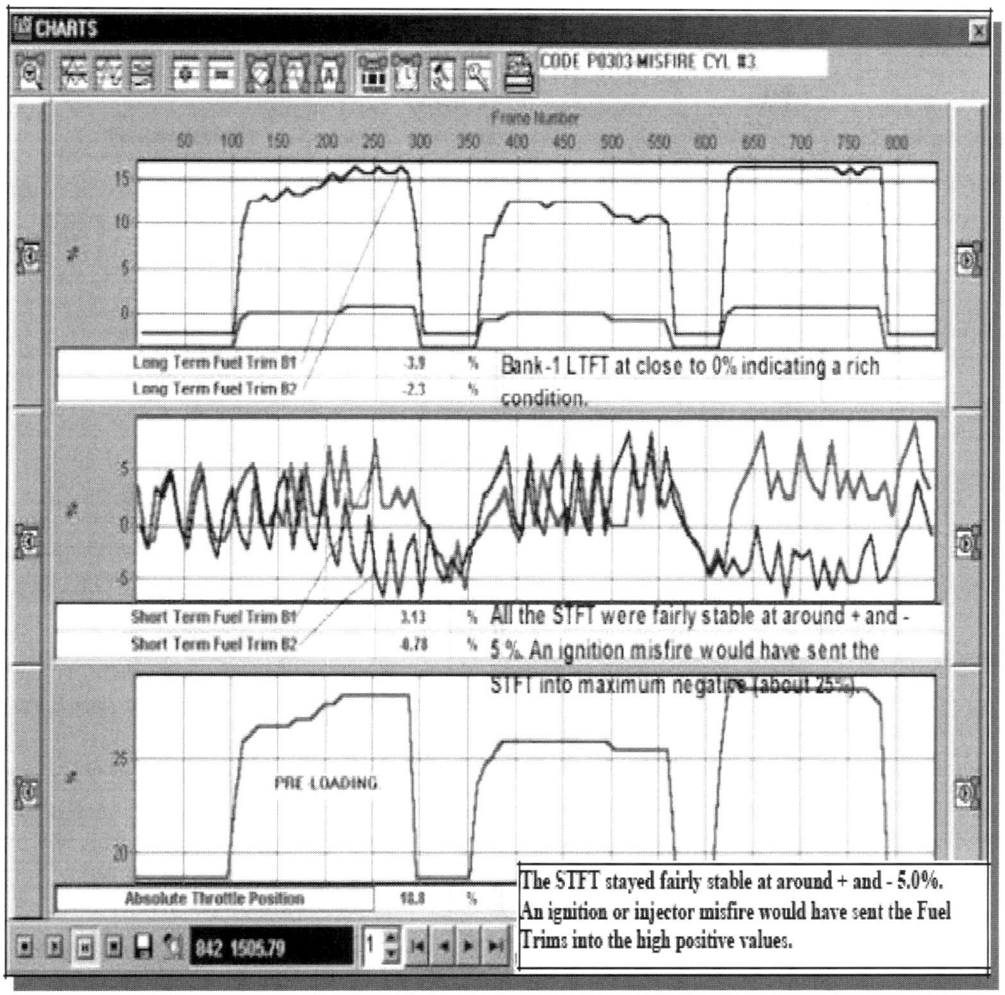

Fig 2 – Chart showing LTFT, STFT and TPS.

The STFT stayed fairly stable at around + and - 5.0%.An ignition or injector misfire would have sent the Fuel Trims into the high positive values.

This case study shows the benefits of using PID scan tool diagnostics. Given today's tightly packed engines, these techniques are very powerful. By simply knowing and analyzing the relation between the different data parameters (PID), a sound diagnostic decision can be reached without reaching inside the engine compartment.

At the very least a good 75% of the diagnostics process can be performed by this means, with a further 25% using whatever manual tests are needed, saving a tremendous amount of time and money.

What I couldn't understand at that time was how could the EGR cause a misfire on only one cylinder or bank-1 (Cyl. #3).Upon further checks it was revealed that the EGR passages toBank-2 were actually clogged. This had the effect of re-routing the EGR gasses toBank-1 only. With such an excessive amount of inert gasses going to Bank-1, cylinder #3 was misfiring. Due to the design of the intake manifold, most of the EGR gasses were only reaching the rear cylinder (Cyl #3), causing a misfire. A thorough manifold cleaning and a new EGR valve solved the problem. A PID graph analysis showed the following after repairs were done.

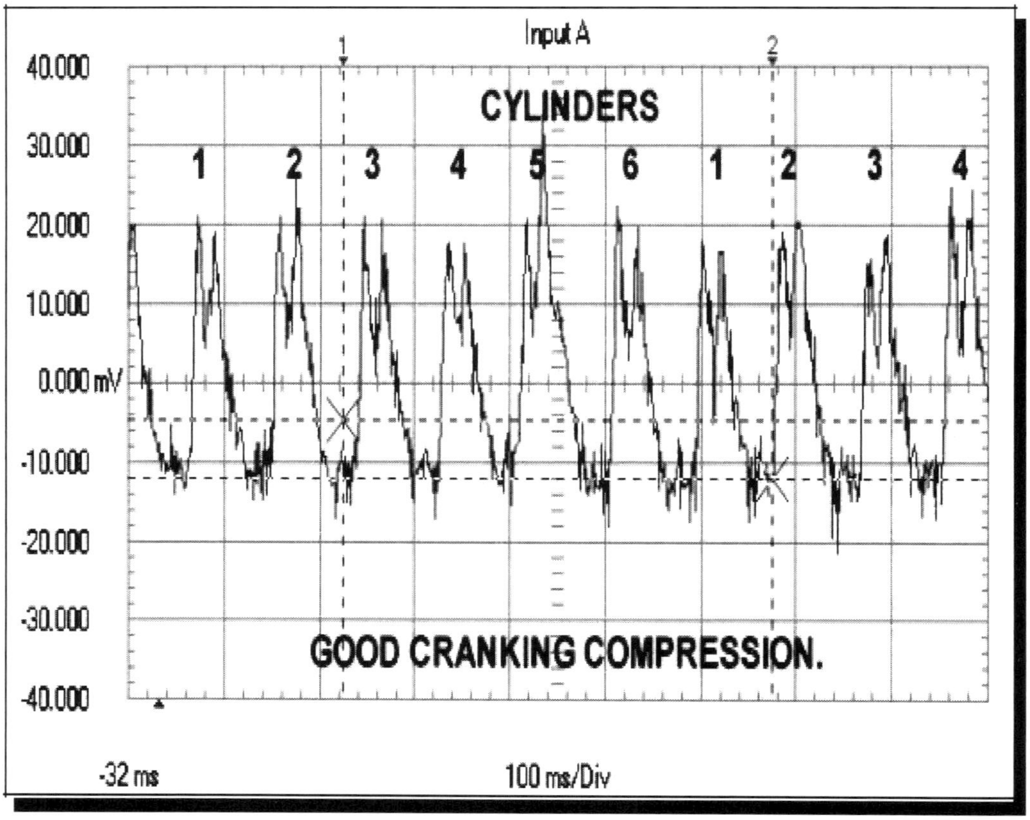

Fig 3 – This high current cranking compression waveform show good mechanical integrity.

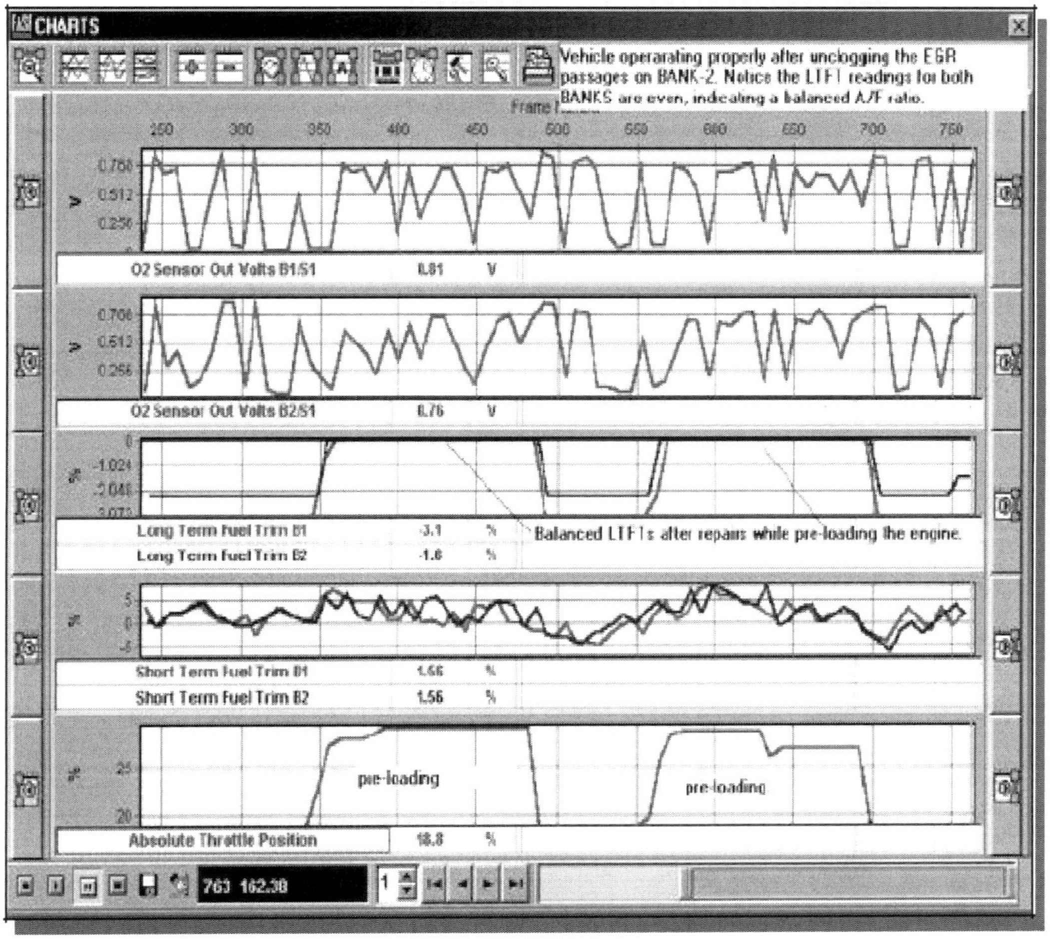

Fig 4 – Scanner PID analysis after repairs were done. Notice the Even LTFT and STFT indicating a balanced A/F ratio.

####

About the Author:

Mandy Concepcion has worked in the automotive field for over 21 years. He holds a Degree in Applied Electronics Engineering as well as an ASE Master & L1 certification. For the past 16 years he has been exclusively involved in the diagnosis of all the different electronic systems found in today's vehicles. It is here where he draws extensive practical knowledge from his experience and hopes to convey it in his books. Mandy also designs and builds his own diagnostic equipment, DVD-Videos and repair software.